DISSECTION PROJECTS

DISSECTION PROJECTS

DAVID WEBSTER

ILLUSTRATED BY ANTOINETTE B. JACKMAN
FRANKLIN WATTS   1988   A FIRST BOOK
NEW YORK   LONDON   TORONTO   SYDNEY

Library of Congress Cataloging-in-Publication Data
Webster, David, 1930-.
Dissection projects/David Webster; illustrated by Antoinette B.
Jackman.
    p.   cm. — (A First book)
Bibliography: p.
Includes index.
Summary: An introduction to dissecting animals, discussing
instruments, techniques, and acquisition of specimens and providing
instructions for the dissection of clams, squids, fish, and others,
as well as specific parts such as a beef heart and calf eye.
ISBN 0-531-10474-5
1. Dissection—Juvenile literature.   [1. Dissection.]
I. Jackman, Antoinette B., ill.   II. Title.
QL812.5W43   1988
591.4'028—dc19   87-25179   CIP   AC

Cover photograph by Laima Druskis.

Diagram 10D on page 45 and diagram on
page 91 by Anne Canevari Green.

All photographs courtesy of Henry Rasof except:
Laima Druskis: pp. 15, 48; Barbara Pfeffer: p. 38.

CONTENTS

DISSECTION PROJECTS

1

DISSECTION: A FUN WAY
TO LEARN ABOUT LIFE

Dissecting is fun. It is also the best way to learn about the structure of animals' bodies, inside and out.

You can cut open a chicken or fish and take out the heart, stomach, and intestine. You can find the strange beaks in a squid's mouth or the muscles inside a lobster's antennae. The lens from the eye of a calf or fish can be used as a little magnifying glass. There is even much to discover by dissecting a chicken leg.

Most of the specimens described in this book can be obtained from a supermarket or fish market, although some can also be obtained from scientific supply companies. You can learn a lot by dissecting some of the foods you buy to eat.

You already have the most important dissecting tools—your hands—but you will still need to assemble a dissection kit with tools to conduct your surgery. You should have a pair of scissors, knives, tweezers, and a magnifying glass. The next chapter tells how to buy and make dissecting instruments. You must remember to handle the sharp tools with care.

This book has detailed instructions for dissecting eleven different animals or animal parts. Of course, many other specimens are just as good to dissect. First, try a few of the dissections described in the book for practice. Then you should be able to do some of your own.

When dissecting, refer to the drawings as well as to the written directions. Italicized terms (for example, *umbo*—the bump on top of a clam shell) in the text are shown in the drawings, and all labeled parts in the drawings are in italics in the text.

It is important to spend time examining the outside appearance of a specimen before you begin to dissect it. You can learn a lot just by touching the specimen and by looking at it carefully. Use the magnifying glass. Compare the structure of the specimen you are studying with similar parts on your own body and the bodies of any pets you may have.

A proper dissection is done only with great patience so that you can uncover and remove body structures without damage. Dissection is not meant to be just a quick hacking job with no time spent on examining and thinking.

You might want to keep some parts of the dissected specimens in a collection. Harder parts, such as bones, teeth, and scales can be dried and glued to display boards. Softer, fleshy parts can be preserved in rubbing alcohol from the drugstore. (Have an adult help you with this. Store alcohol in a safe place away from small children). Put the parts in separate glass jars with labels for identification. Specimens also can be preserved in a freezer before or after dissecting. Your dissection collection might make a good science fair project for school.

Perhaps the only thing you will not like about dissecting is the cleanup after you are done. It is not fair to leave a mess for someone else to clean. Put down newspapers first so that nothing will get on the desk or table where you are working. Outside is often a good place to work. When you are done, wash your instruments in soap and water; then dry them with paper towels.

After you have had some practice, you may want to try more difficult dissections, for example, those of preserved specimens. Ask the biology teacher in your high school for the addresses of scientific supply companies that sell such specimens as sharks, frogs, pigeons, rats, and fetal pigs.

Always dissect with the purpose of learning about the strange and complex design of nature. The more you know, the greater respect you will have for life—your own and that of all the other splendid creatures on earth.

2

DISSECTION INSTRUMENTS

In order to dissect properly, you will need dissection tools. You might be able to buy a dissection set at a toy store that sells science kits. However, you can assemble your own set by making some instruments and buying others.

1. Dissecting scissors/Scissors. The most important dissection tool is a good pair of scissors. The scissors used in schools for cutting paper are usually not sharp enough to cut flesh. Other desk scissors you may have at home probably will be too large. The points on scissors for dissection should be about 2 to 3 inches (5 to 8 cm) long. The cutting edges should be sharp and should shear together when closing. You can purchase a good pair of scissors at a drugstore or fabric store. Drugstores sell sharp scissors for cutting bandages. Ask for embroidery scissors at a fabric store. **Be careful when using these scissors.**

2. Scalpel/Knife. A scalpel has a short, sharp blade attached to a strong handle. Craft supply stores sell Exacto knives used to cut balsa wood for model making. The blade of a craft knife is very sharp and can be replaced with a new blade when dull. The longer blade of a kitchen paring knife will be useful for removing larger parts from certain specimens. You also could get a few

Scientific name of tool	Common name of tool	Use in dissecting
1. Dissecting scissors	Scissors	Cutting
2. Scalpel	Knife	Cutting and making small holes
3. Forceps	Tweezers	Picking up small things
4. Teasing needle	Needle	Separating small parts
5. Pipette	Dropper	Picking up liquids
6. Magnifying glass/microscope	Magnifying glass/microscope	Looking at small parts
7. Dissection tray	Tray/pan	Holding specimens when dissecting

single-edge razor blades to use for slicing into tough tissues. **Always work under adult supervision when using these tools.**

3. Forceps/Tweezers. Drugstores sell tweezers for taking out splinters. The tips of the forceps should be pointed rather than blunt. Use the tweezers with care so that the tips do not become bent.

4. Teasing needle/Needle. Sometimes teasing means to bother or annoy someone on purpose. In dissection, teasing means to separate delicate parts from the specimen. A teasing needle should have a rounded wooden handle. It is easy to make one from a pencil and a long sewing needle. Cut off the point of the pencil and tape on the needle so that it sticks out several inches.

5. *Pipette/Dropper.* A pipette, which is just a simple eyedropper, can be bought in almost any drugstore. The pipette is useful for picking up liquids and for dropping them on microscope slides for examination.

6. *Magnifying glass/Microscope.* A magnifying glass should be easy to get. A "glass' with a plastic lens works fine and is much cheaper than one with a glass lens. When using a magnifying glass, hold the lens as close to your eye as you can. To focus, move the object that you want to see back and forth until it is clear. Be sure the specimen is well lighted. You need a microscope to see small details.

7. *Dissection tray/Pan.* An aluminum cookie sheet makes a good dissection tray. The low sides prevent juices from damaging the

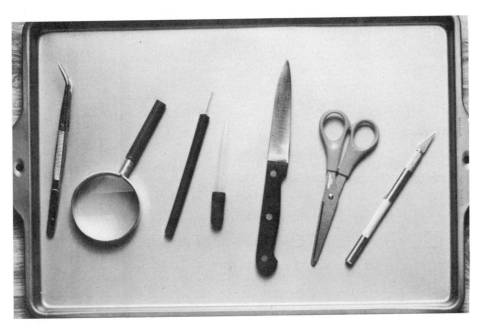

A homemade dissection set

surface on which you work. The pan should be about 10 inches by 15 inches (25 by 38 cm).

A pair of pliers or a nutcracker will also come in handy in some of the activities.

Cleanup. Good dissecting tools require good care. Do not use the sharp instruments to cut thick bones or other hard materials. Thoroughly wash and dry all of the tools as you finish using them. If you live with younger brothers or sisters, keep the tools in a safe place so that no one will damage anything or hurt themselves.

3

WHERE TO
GET SPECIMENS

Specimens for dissecting can be obtained from many different sources.

1. Fish market. Perhaps the best source of specimens is a fish market. You will need a "whole" fish rather than a "cleaned" one that has had its internal organs removed. Fish stores sell other kinds of animals, too. Lobsters, crabs and squid make interesting dissection specimens. Shellfish such as clams, oysters, and mussels should also be available. The kinds of seafoods sold vary in different seasons and in different parts of the country.

2. Supermarket. Chicken parts from a supermarket might be good specimens for your first dissection. Packages of cut-up bones are sold for making soup and to give to dogs. Ask the butcher to save you a few larger bones. Other possibilities are turkey parts, duck and pig knuckles.

3. Ethnic market. An ethnic market is one that sells special foods for people who still eat the foods found in their native countries. Many Italians, for example, like to buy whole chickens with the heads and feet still attached. Many people of French extraction enjoy escargot (large land snails); Greek-Americans like octopus.

Many recent immigrants or people with strong ethnic ties live in distinct neighborhoods or concentrate in cities or districts. Stores in these sections may carry items perfect for special dissections.

4. Slaughterhouse. Perhaps there is a slaughterhouse nearby, where steers, calves, sheep, or hogs are killed for meat. You can find out by looking under "Abattoir" in the Yellow Pages of the telephone book. Call up first to see what animal parts might be available. They should have hearts, lungs, brains, eyes, and stomachs. There probably will be a small charge for these parts. If there are no local slaughterhouses, you could visit one when you travel during vacations or in the summertime. The parts can be kept frozen until you want to dissect them.

5. Science classroom. In school, biology classes dissect preserved animals. Most frequently studied are worms, frogs, white rats, and fetal pigs. Ask your own teacher how you could talk to a science teacher in the junior high school or high school. There is a good chance that he or she would be happy to give you something to dissect.

6. Scientific supply company. A number of special companies sell preserved specimens such as starfish, frogs, squid, fish, clams, crayfish, and earthworms. Usually preserved specimens are used in high school biology classes, but if you can't get some of the specimens described in this book from sources 1 through 4 or 6, you might want to use the preserved kind. A teacher can help you order such specimens.

An ethnic market is a good source of dissection specimens.

*You can also catch your own
fish for dissections.*

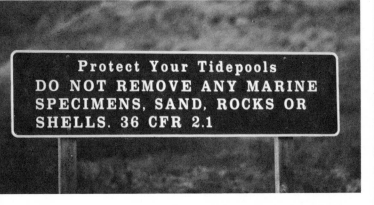

Protect Your Tidepools
DO NOT REMOVE ANY MARINE
SPECIMENS, SAND, ROCKS OR
SHELLS. 36 CFR 2.1

*The no-collecting signs at this
tidepool area in Point Loma, California,
are in several languages.*

7. The Great Outdoors. If you like to fish, you should be able to catch a fish that you could dissect. You may be able to collect such creatures as starfish, sea urchins, crabs, mussels, and eels. Make sure there are no restrictions on these creatures. In California, for example, shellfish may be poisonous during certain months of the year. In addition, special beaches may prohibit the collection of tidepool animals. Signs are often posted wth retrictions and warnings. Insects also made good specimens. Dissecting them is made easier if you use a magnifying lens.

Clams

DISSECTING
A CLAM

A clam is a good specimen for your first dissection. The kind of clam you can get will depend on where you live. Fish stores along the East Coast sell live quahogs, steamers, and mussels. Similar kinds of shellfish can be purchased on the West Coast. (Scientific supply companies also sell clams.) Any kind of clam you dissect will have about the same structure as the quahog (pronounced Ko-hog) described here. Quahogs and other clams can be kept alive for many weeks in your refrigerator.

The quahog is a member of a group of animals called mollusks. Mollusks are simple animals with soft bodies usually enclosed in a protective shell made from chemicals obtained from the water. Mollusks with a pair of shells, such as clams, scallops, and mussels, are called bivalves (*bi* means "two"). Snails and conches are mollusks with only a single coiled shell. The garden slug and octopus are mollusks with no shell at all.

EXTERNAL EXAMINATION

Before dissecting your quahog, look at the outside of its shell (Figure 1). Notice that the shell is marked with a series of curved lines. These rings show how the shell changed size as the mol-

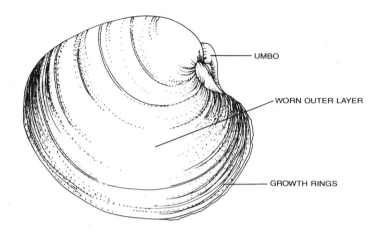

UMBO

WORN OUTER LAYER

GROWTH RINGS

Figure 1. Outside of a quahog shell

lusk grew older. The oldest part of the shell is the bump at the top called the *umbo*. Look at the umbo with your magnifying glass. The tiny cap at the tip was the baby quahog's first shell.

How many *growth rings* can you count on your shell? Unlike a tree, the age of a mollusk cannot be measured by counting its shell rings.

The shell is made up of three layers: the outer layer, the middle layer, and the pearly layer inside. The *outer layer* is composed of a dark, horny material. This hard covering protects the middle layer from being dissolved by the weak acids in seawater. The outer layer often gets worn away from the older parts of the shell. Where on your shell has the outer layer rubbed off?

OPENING THE CLAM

Clams hold their shells closed with two powerful muscles. Try to pull apart the shells with your hands. The little quahog is stronger than you are!

Clammers open a quahog shell with a special short knife. The knife blade is forced between the shells so that the muscles can be cut. You could try this with a paring knife, but you might cut your hand or break the knife. Try the following instead:

1. Place the quahog in a pan of very hot tap water. This relaxes the quahog's muscles.
2. Leave the quahog in the water for 10 to 20 minutes, until the shells begin to open.
3. Cut the *shell muscles* with a paring knife (see Figure 2). **Be careful.** Better yet: have an adult help you.
4. Force the shells apart with your hands until the shells are spread open like the pages of a book.

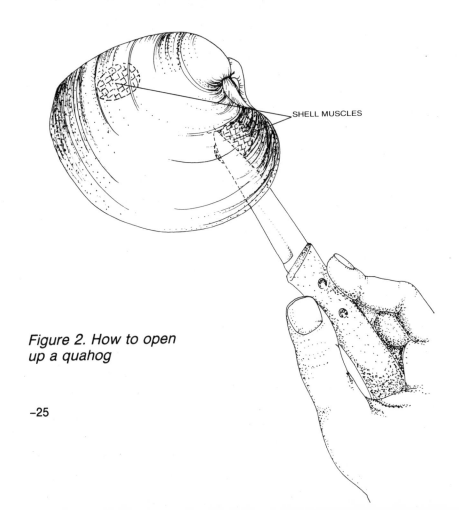

SHELL MUSCLES

Figure 2. How to open up a quahog

INTERNAL STRUCTURE

Refer to Figure 3 for your examination of the quahog's inner structure. The shell muscles—the *adductor muscles*—that you cut will still be stuck to the shells. Muscle tissue is usually reddish in color, so the adductor muscles will appear as pinkish disks attached near the edges of the shell.

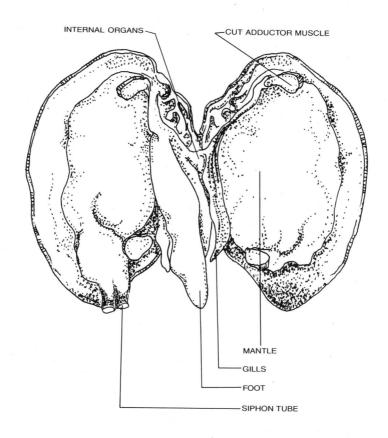

INTERNAL ORGANS

CUT ADDUCTOR MUSCLE

MANTLE

GILLS

FOOT

SIPHON TUBE

Figure 3. Internal organs of a quahog

Sticking up between the two shells you will see the clam's fleshy *foot*, the biggest part of its body. Squeeze the foot and note its shape. The pointed tip is used as a shovel for digging. A clam moves by extending its foot far out of the shell into the sand. Then the tip of the foot swells with blood and acts as an anchor. As the foot is withdrawn into the shell, the clam is pulled slowly forward.

You should find the sheetlike *gills* hanging from the base of the foot. There are two gills folded together on each side. Use your forceps to separate the two gills. If you have a microscope, snip off a piece of gill and look at it under the low power. Can you see tiny holes or pores on the surface? The gills serve as both a breathing organ and a feeding organ. As water circulates through microscopic pores in the gills, food particles of tiny plants and animals are trapped on the surface. The food gradually is moved forward into the clam's mouth.

Water is drawn into the clam through a short *siphon tube* and forced out through a second tube. Look for the siphon tubes near the more pointed side of the shell. The ends of the tubes can be recognized by their dark brown color. Try poking the tip of your pipette through one of the tubes.

The inside of the shell is covered by a thin skin called the *mantle*. Chemicals secreted by the edge of the mantle harden to form the growing shell. See if you can remove the mantle in a single piece from one side. Pull it away from the shell with your fingers while using the scissors to cut places where it is connected along the edges.

The *internal organs* are inside the top part of the foot. They include the stomach, liver, heart, kidney, and anus. The intestine coils back and forth inside the foot. Cut open the foot at the top and examine its internal structure. It will be difficult to identify any of the organs listed above.

A quahog has no brain, backbone, or eyes. What other body parts that *you* have are absent in the quahog?

INSIDE OF SHELL

To examine the inside of the shell (Figure 4), remove all of the soft parts with the scalpel. (**Be careful.**) Scrape away the adductor

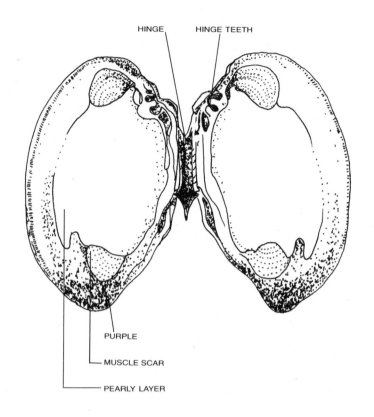

HINGE HINGE TEETH

PURPLE

MUSCLE SCAR

PEARLY LAYER

Figure 4. Inside of a quahog shell

muscles where they attach to the shell. The shiny spots under the muscles are known as *muscle scars.*

 The *pearly layer* on the inside of the shell is composed of the same substance used by oysters in making pearls. Atlantic Coast Indians made beads from quahog shells. Beads made from the small *purple* area on the shell's rim were more valuable.

 The two shells are attached at the top with a flexible *hinge* composed of tough cartilage. When the adductor muscles relax, the cartilage hinge springs the shells apart. Squeeze the shells together and watch them pop apart as you let go.

Inside the umbo you can see the *hinge teeth.* These interlocking grooves and ridges line up the shells when they shut. Notice how the hinge teeth mesh together as you close the shells. The hinge is quite strong. Force back on the shells and see if you can crack the hinge and separate the shells.

Take one shell and try to break it with your bare hands. Can you do it? If not, put the shell on a workbench and carefully hit it with a hammer. Look at a broken edge with the magnifying glass. Two layers will show: the middle layer and the pearly inner layer; the horny outer layer is too thin to see.

Project: Dissolving a Clam Shell
Drop a piece of clam shell into a glass of vinegar. Bubbles will start forming immediately as the acid in the vinegar reacts with the calcium in the shell to make carbon dioxide gas. Put the shell in the vinegar aside and look at it occasionally for several days. How long does it continue to fizz? Compare shells from different types of clams.

When the fizzing stops, take out the shell and examine it. Does it seem to be any smaller? Can you break it?

You can use a balloon to measure how much gas is produced. Get a soda bottle and break up several shells into small pieces. Fill the bottle with 3 inches (almost 8 cm) of vinegar, drop in the crushed shells, and place the balloon over the top of the bottle. What happens to the balloon?

EATING QUAHOGS

Thin-shelled clams are eaten by animals such as lobsters and sea otters. The thickness of the quahog's shell protects it from almost all natural predators. People are quahogs' only real enemy.

Large quahogs are the main ingredient of clam chowder. Every part of the mollusk except the shell is chopped up and cooked in a milk broth. Medium-size quahogs, called cherry-stones, are used for baked stuffed clams. The smallest quahogs, known as little necks, are eaten raw on the half shell. Have you ever eaten raw clams or oysters? Do not eat any of your specimens unless you have the permission of a teacher or parent.

5

DISSECTING
A SQUID

You would not think that the squid is related to the clam. The squid's soft body is not protected by an outside shell. Unlike the sluggish clam, the squid's best defense is its ability to leave the scene of danger in a hurry. In spite of some obvious differences, however, clams and squids are both mollusks. The squid does have a thin, frail shell, which is embedded inside its body. The octopus, another mollusk, has no shell at all.

Giant squids are the largest animals without backbones. Sometimes these deep-sea "monsters" reach a length of over 50 feet (15 m). Occasionally, a giant squid will attack a whale, the world's largest animal *with* a backbone. Round scars, known as sucker marks, have been found on the skins of whales.

WHERE TO GET SQUID

Fish markets throughout the country sell squid. Along the Atlantic and Pacific coasts, fresh squid can be purchased during most of the year. Frozen squid is available in many inland areas. Scientific supply companies also sell squid. Be sure the squid you get is whole rather than cleaned. The internal organs will have been removed from cleaned squid. You want a whole squid that has all the parts still inside.

-31

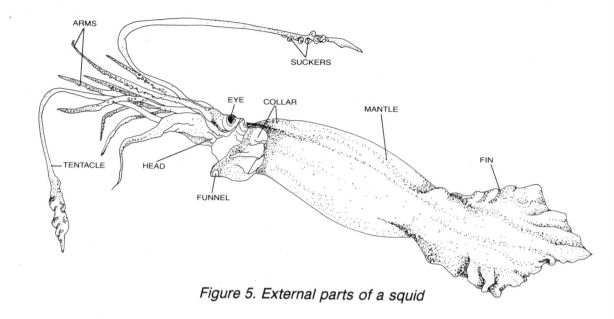

Figure 5. External parts of a squid

You can preserve fresh squid in your freezer for as long as you wish. A frozen squid can be thawed out at room temperature in several hours or in hot water in about fifteen minutes.

EXTERNAL EXAMINATION

Refer to Figure 5 when making your external examination of the squid.

A squid is a strange-looking creature. The main part of its body is the thick and muscular skin called the *mantle*. The pointed end of the mantle has a pair of triangular folds or *fins*. The fins can be flapped up and down to propel the animal slowly or to change its direction of movement.

Spread out the fins of your squid so that you can see their shape. Do the squid's fins remind you of the fins on the tail of a rocket? Notice that one side of the fin is speckled with purplish-red spots. How do you think the spots help the squid? Look at the spots with your magnifying glass. Which spots are darker, the larger ones or the smaller ones? What color is the other side of the fin?

The edge of the mantle toward the *head* ends in the *collar*. Poke a pencil through the collar and underneath the mantle to see how far the space extends toward the fins. The squid's head consists mainly of the two large *eyes* and a mouth surrounded by many skinny *arms*. How many arms does the squid have? How does this number compare with the arms of an octopus? How many longer arms are there?

When a squid swims, the arms are pressed together and aid in steering. The two longer arms, the *tentacles*, are used to reach out and seize prey; the shorter arms hold the prey near the mouth during eating. The *suckers* on the arms help the squid to hold on. Which arms have suction cups along their entire length? Look at one cup with a magnifying glass. Can you see the toothed ring that lines the inside of the cup? Some of the rings may have fallen out of the suckers. Use your forceps to remove a tooth ring. Place it on a microscope slide and view it under low power. You might think you are looking at a miniature glass crown.

Underneath the eyes you should find the *funnel*. Push the handle of your teasing needle through the opening in the funnel. There is a larger opening at the other end of the funnel opposite the hole. In what way is the squid's funnel the same as a real funnel? When water is forced out of the funnel hole, the squid is moved rapidly by jet propulsion. This is the animal's usual method of escape. The squid can change its direction by moving the funnel.

INTERNAL ORGANS

If you plan to cook and eat any of the squid, sterilize your dissection tools by boiling in water for ten minutes before you begin.

Perhaps the most interesting part inside the squid is its shell. It is a long, feather-shaped structure that looks like a sliver of clear plastic. To remove the *shell* (Figure 6), stretch out the squid with the colored surfaces of the fins facing upward. You should be able to feel the hard tip of the shell just under a pointed bump at the rim of the collar. Use your dissection scissors to cut open the skin covering the end of the shell, and then pull it out with the forceps.

Above: *squid (calamari in Italian)
and octopus (pulpo in Italian)*

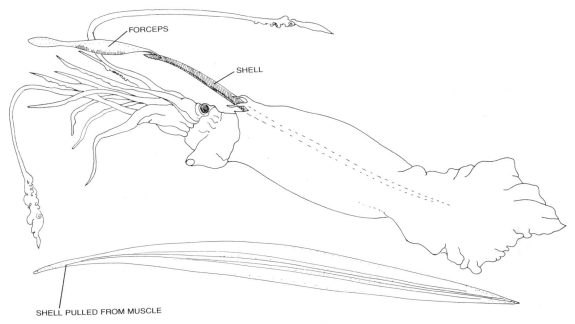

FORCEPS

SHELL

SHELL PULLED FROM MUSCLE

*Figure 6. How to remove
the squid's shell*

The shell should be almost as long as the mantle, but it may
have been broken into pieces when the dead squid was handled
at the fish market. If this has happened, cut a slit down the mantle
with the scissors until you uncover the rest of the shell. Keep the
shell to show to your friends. See if they can guess what it is.

Cut the *mantle* open all the way past the fins (see Figure 7).
**Be sure to work under adult supervision when using the knife or
scalpel.** As you cut, lift up the mantle with your forceps so that
you do not damage the organs underneath. Note the two feathery
gills used to remove oxygen from the seawater. You should be
able to see the tiny blood vessels of the gills with your magnifying
glass. The thin sac between the gills is the *stomach pouch*. Cut
open the stomach and examine its contents. Can you tell what
the squid has eaten?

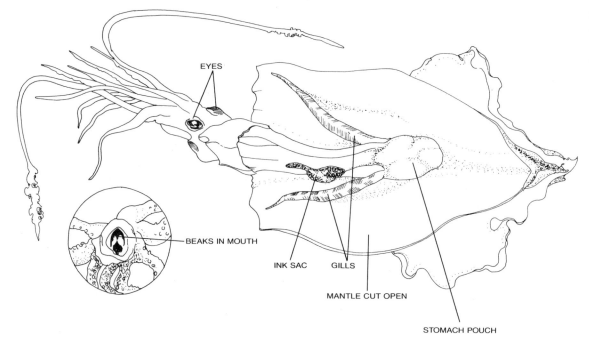

EYES

BEAKS IN MOUTH

INK SAC   GILLS

MANTLE CUT OPEN

STOMACH POUCH

Figure 7. Internal organs of a squid

The *ink sac* is the silvery pouch near the stomach. Puncture the sack with the point of your scalpel to release the "ink." You can write your name with the squid's ink. Use the pointed end of the squid's shell as an old-fashioned quill pen. Squid ink was once used to make commercial ink.

When attacked, the squid sometimes squirts out a cloud of inky liquid. Some scientists think the animal uses the dark water as a "smoke screen" in which to hide. Others believe that the ink serves to distract the enemy while the squid swims off in another direction.

Spread apart the arms so that you can find the squid's mouth in the center. Instead of teeth, the squid has a pair of curved *beaks* shaped like a parrot's bill. The strong jaws kill prey and bite off large chunks of flesh, which are swallowed without chewing. Squids eat small fish, crabs, and even other squid. Snip the skin around the mouth and remove the beaks. In what way are the beaks like the shell? Wash off the two beaks and let them dry out for display.

The squid has well-developed *eyes*. Their structure is surprisingly similar to the design of your own eyes. Try to remove an eyeball by carefully cutting it away from the soft skin. If the scissors cut into the eyeball, it will collapse as the watery eye fluid oozes out. You can collect some of the fluid in your pipette.

There is a lens inside the eyeball. It is a clear sphere about ⅛ inch (3 mm) in diameter. Pick up the lens with the forceps and place it on a piece of paper. Did you notice that the lens is slightly flattened on one side? Put the lens on top of some small print from a newspaper. The lens will act like a miniature magnifying glass and enlarge the letters underneath.

The squid does have a brain, but it is too tiny for you to locate.

FRIED SQUID!

Squid is considered a delicacy by many people in countries surrounding the Mediterranean Sea. Its taste resembles that of chicken. In this country, squid is sometimes listed on the menu in fancy restaurants under the Italian name of calamari. The cleaned mantle and arms are steamed in a tomato sauce or deep-fried with a batter coating. Maybe you can persuade your mother or father to help you (and your friends) prepare some squid to eat. Don't overcook it, or it will become rubbery.

DISSECTING
A LOBSTER OR
CRAYFISH

Most people know lobsters only as a tasty delicacy. Have you ever eaten one? If so, you probably learned something about a lobster's body as you dissected it during your meal.

A lobster is a lot different from a clam or a squid. In spite of its hard external skeleton, the lobster can move quickly. Attached to all parts of the lobster's body are a variety of appendages such as legs, claws, and antennae. Each kind of appendage has a special function.

Live lobsters can be purchased in fish markets in most cities. Maine lobsters flown from the New England coast can often be found in large saltwater tanks in fish markets.

Many people think crayfish are baby lobsters, but there are several important differences between crayfish and lobsters. Crayfish never grow much larger than 6 inches (15 cm) long whereas lobsters have been found as heavy as 35 pounds (16 kg), and occasionally even heavier. Also, crayfish live only in fresh water, whereas lobsters live in salty ocean water.

Weighing lobsters
in Maine

In every other way, however, the body of a crayfish is almost identical to a lobster's. You could dissect a crayfish by following the same directions as for a lobster.

Crayfish are eaten in certain parts of the South, especially around New Orleans. Some pet shops (and scientific supply companies) sell crayfish, too. Since crayfish live in ponds and streams throughout the United States, you might be able to find one yourself. Look under rocks in a stream or on the muddy bottom of a pond or swamp.

Project: Keeping a Pet Crayfish

If you are able to get a live crayfish, keep it for a while as a pet. A 10-gallon (38-liter) aquarium will make a good home for your crayfish. Fill the tank with 2 inches (5 cm) of water and put in several rocks for hiding places. Feed the crayfish bits of bologna once a week.

You can pick up your crayfish by holding its body just behind the places where the claws attach. Examine the crayfish with a magnifying glass while you hold it out of water. Can you see the eye stalks or mouth parts moving?

Hold a pencil in front of the crayfish near one of the claws. When the crayfish bites the pencil, let go of the crayfish so that it is hanging from the pencil. How long can it hold on? Do not try to pull the pencil away or the claw might be ripped off.

If you keep the crayfish long enough, it might shed its shell, or *molt*. After molting, the crayfish's new shell is soft for several days. Feel it.

OBSERVING A LIVE LOBSTER

Put several inches (or centimeters) of fresh water into a bathtub or large sink. Drop in the lobster and watch as it walks around underwater. In which direction does it usually move? How are the antennae used? Turn the lobster upside down in the water. How does it try to get right side up?

What color is a live lobster? What part of the lobster is reddish-orange? Where is the shell bluish?

COOKING THE LOBSTER

The best way to prepare a lobster for dissection is to cook it in boiling water. Have an adult or older brother or sister help you. Here is an easy method:

1. Get a pot or metal bucket large enough to hold 4 to 6 quarts (or liters) of water.
2. Heat the water on your stove until the water boils.
3. Drop in the lobster and cook it for 10 minutes.
4. Cool the lobster in cold water.

EXTERNAL EXAMINATION

Refer to Figure 8 when making the external examination of the lobster. Probably the first thing you notice about the cooked lobster is that it has turned all red. A hard shell, or *cuticle*, covers all

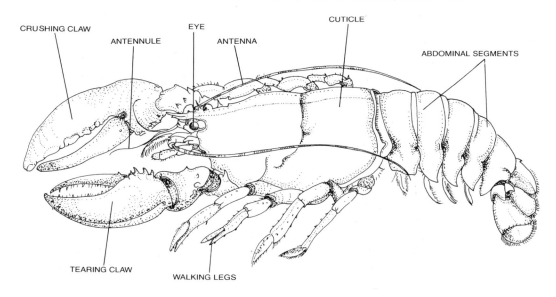

Figure 8. External parts of a lobster

parts of the lobster's body, including the appendages. An outside skeleton has the obvious advantage of protection. But since the cuticle is rigid, the appendages must have many joints to permit movement. The stiff parts between the joints are known as segments. In Figure 8 the *abdominal segments* have been labeled.

A lobster has two main body parts, which most people call the body and the tail. However, the body is really a combination of the head and thorax (chest) and has a big name: cephalothorax. Biologists call the tail of the lobster the abdomen.

Use your magnifying glass to look at all the joints in an *antenna*. There are more than 250 segments in each long antenna to give it complete flexibility. The shorter branched antennae are called *antennules*. In addition to feeling, the lobster uses its antennae and antennules for tasting.

The *eyes* are mounted on the ends of movable stalks, enabling the lobster to see in all directions at the same time. The lobster's eye is not designed like your own, with a lens to focus images. Instead, the eye has thousands of tiny sections, which help the animal detect motion.

The mouth is in front beneath the bases of the antennae. A variety of external parts break up food and pass it on to the mouth. Use your teasing needle to move and separate the mouth parts.

Look at a claw. Can you count the six joints? Move the claw in different directions and watch how each part allows movement in only a single direction. The three joints of your arm are in your wrist, elbow, and shoulder. Even with six joints, can a lobster use its claw to rub its stomach or scratch its back as you can with your hand?

The two claws are not the same; one is larger than the other. The bigger, *crushing claw* is used, as the name indicates, for crushing. The *tearing claw*, smaller and thinner, is used for grabbing and tearing soft prey. Both claws serve as awesome defense weapons as well. Which claw has sharper teeth? What part of the claw has bristles?

How many *walking legs* does the lobster have? Which ones have no claws? How might the appendages under the abdomen be used?

-42

DISSECTING THE CLAWS

There is no reason why you cannot eat the edible parts of the lobster after the dissection. If you want to do this, however, you must sterilize your dissection tools before using them. Place them in boiling water for 10 minutes before starting. Do not eat any of the lobster without getting permission from an adult.

The directions for dissecting a lobster are almost identical to the procedures used when eating one. Most people like to start with the claws, since they contain a lot of tasty meat. To dissect the claw (see Figure 9):

1. Bend the claw down and back until it breaks off at the base.
2. Snap off the main part of the claw from the three smaller sections.
3. Use pliers or a nutcracker to crack open the shell on the main part of the claw. Do not break the shell from the pincer.
4. Separate the meat from the base of the pincer with a scalpel or teasing needle.

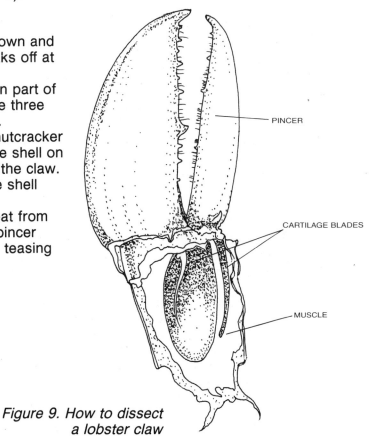

PINCER

CARTILAGE BLADES

MUSCLE

Figure 9. How to dissect a lobster claw

If you have trouble with the first claw, try again on the other one. Look for the thin *cartilage blades* attached to the pincer. The *muscles* pull on the blades when the *pincer* is moved. Why is one cartilage blade so much larger than the other?

The meat in the claw makes up the muscles used to open and close the pincer. Large claw muscles are needed to provide the strength for defense and to crush prey. There is a little meat in the other claw segments. Crack the shell with the pliers and remove the muscle with the forceps. Try eating some of the claw meat after it has been dipped into melted butter.

THE ABDOMEN

The lobster's abdomen, or tail, contains more meat than the claws. Since the meat is muscle, this means that the abdomen is capable of powerful movements. By snapping and curling its tail, the animal can shoot backward in rapid jerks.

To dissect the abdomen,

1. Break the tail from the body (see Figure 10A).
2. Snap off the tail fins (see Figure 10B).
3. Push out the meat with your thumb (see Figure 10C).
4. Separate the tail meat (see Figure 10D).

Before the tail meat is eaten, it is deveined. Deveining is the polite way of describing the removal of the intestine containing digested food. The thin, dark tube inside the muscles is the intestine; remove it with your forceps or the point of the scalpel.

Examine the abdominal shell. Bend it up and down so that you can see how the five flexible joints move. Look inside. Notice how the shells of the six segments overlap when the tail is straightened. The tail shell can be washed out and saved.

The legs have thin muscles inside the shells. To eat this meat, snap off a leg at the base and put the open end into your mouth. Then squeeze the leg from the outer end to force the muscle into your mouth.

There is an even thinner muscle inside the antenna. See if you can remove some of it with your dissection tools.

-44

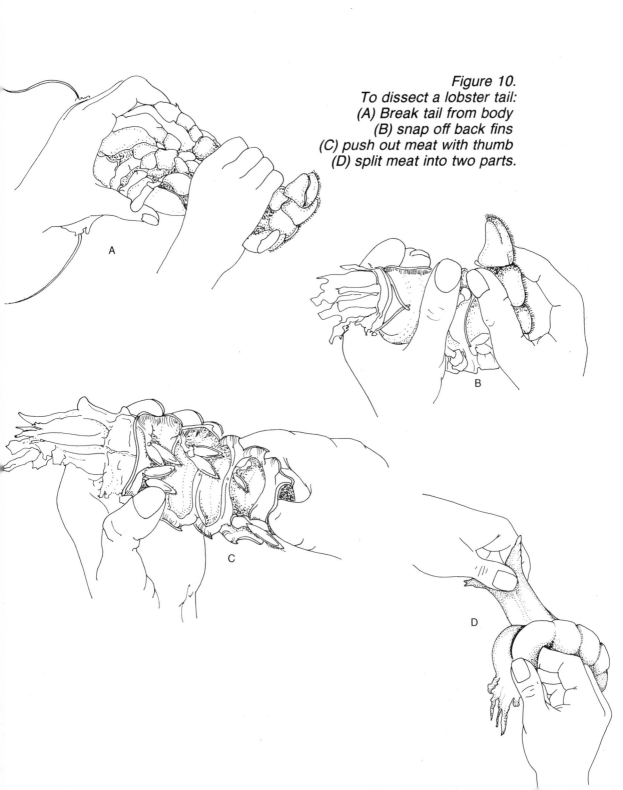

Figure 10.
To dissect a lobster tail:
(A) Break tail from body
(B) snap off back fins
(C) push out meat with thumb
(D) split meat into two parts.

INTERNAL ORGANS

The final step in this dissection is to open up the lobster's body (see Figure 11). To do this, grasp the legs with one hand and the back edge of the shell with the other hand. Lift the shell up and forward until it separates. Some of the internal organs will be attached to the inside of the *shell*. The greenish paste is the remains of the *liver*. It is called tomalley and can be eaten. Sometimes you find pink mush inside. These are eggs, or coral, which also can be eaten.

Figure 11. Internal organs of a lobster

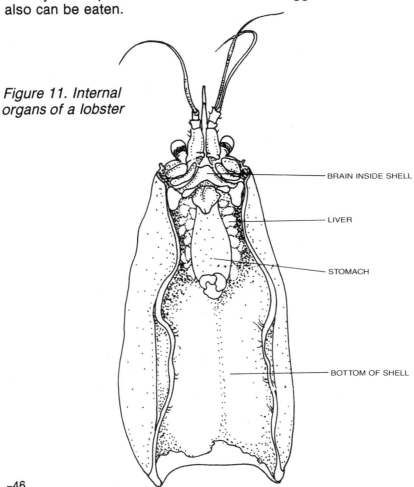

BRAIN INSIDE SHELL

LIVER

STOMACH

BOTTOM OF SHELL

-46

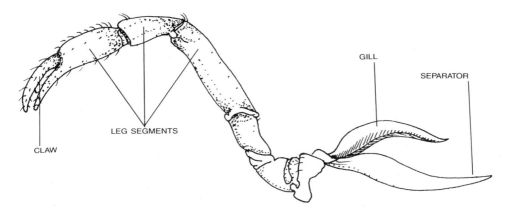

Figure 12. Lobster leg with gill attached

The *stomach* is the bulge at the front. Cut it open and look inside. You might find the spines from a sea urchin that the lobster had for its last meal. Part of the stomach is lined with hard teeth for pulverizing tough food. The tiny *brain* is located between the eyes. Can you find it?

Now look at the lower part of the body with the legs (see Figure 12). The feather structures on both sides are the *gills*. These are sometimes called "deadmen." Do not eat them. A gill is attached to the base of each leg. Try to remove a leg with the gills attached. Beside the gill is a *separator*. Since the gills are attached to the legs, the gills move back and forth when the lobster walks or swims. This stirs up the water under the shell and permits more absorption of oxygen from the seawater.

How many *leg segments* are there? Is there a *claw* on each leg? There are small morsels of meat where the legs attach to the body. Snap off the legs and pick out the "body meat" with the teasing needle. Some people break the body in half lengthwise to make it easier to remove the muscles that move the legs.

The lobster does have a heart, but it is too small for you to find.

Did you enjoy dissecting and eating your lobster?

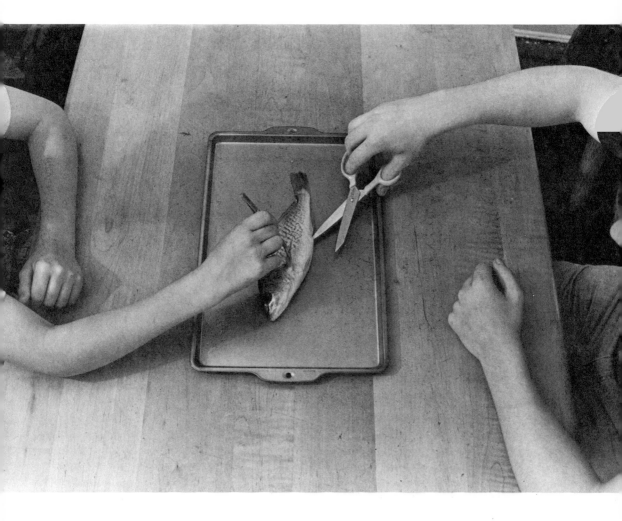

Dissecting with a friend

7

DISSECTING
A FISH

It should not be too hard for you to get a fish for dissecting. You will need a whole fish rather than a cleaned one. Some kinds of whole fish that you might be able to buy are cod, red snapper, herring, ocean perch, whiting, and mackerel. Since fish are usually not expensive, try to get a big one so that the parts are easier to find.

The dissection directions that follow pertain to the mackerel, a common fish of the Atlantic Ocean in the Northeast. However, all fish have basic similarities, so these instructions will apply to whatever kind of fish you use. In fact, an interesting project would be to dissect different types of fish and compare their internal and external structures, why some are more aggressive, and so on.

EXTERNAL EXAMINATION

What do you notice about the mackerel's coloration? The coloring of many fish helps them to be camouflaged in their underwater surroundings. The dark mottling on top of the mackerel causes it to blend in with the dark bottom of the ocean. How does the light color of the mackerel's stomach serve as camouflage when the fish is near the surface?

-49

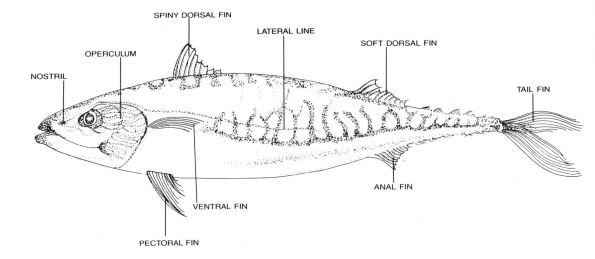

SPINY DORSAL FIN

LATERAL LINE

SOFT DORSAL FIN

OPERCULUM

NOSTRIL

TAIL FIN

ANAL FIN

VENTRAL FIN

PECTORAL FIN

Figure 13. External parts of a mackerel

Refer to Figure 13 during the following examination of your fish.

How many fins does the mackerel have? Count them. You may have to pull up some fins in order to see them better. What is the proper name for the side fin? Where are the *dorsal fins*? Which fin is the largest? Which fin has sharp spines for protection?

Fish have a sense organ unlike any that you have—the *lateral line*, which detects vibrations and changes in water pressure. Find the lateral lines on both sides of your fish.

Open the mouth of the fish with your fingers and look inside. What color is the tongue? How do mackerel teeth differ from your own? Do you think the mackerel chews its food or swallows it whole? Can you find more teeth partway down the throat? Above the mouth are two *nostrils*, which the fish uses to detect chemicals in the water.

Lift a gill cover, or *operculum*, so that you can see the gills underneath. Do the gills remind you of red plastic feathers? How many gills are on each side? Cut out a gill and examine it with a magnifying glass or microscope. Pull out the gill covers and look up into the mouth. Why are there openings from the mouth into the gills?

The body of the fish is covered by thousands of overlapping scales. The biggest scales are on the top side. Scales protect against injury while allowing flexible body movements. Lift off a scale with your scalpel and look at it with a magnifying glass or microscope. The lines on the scale are formed as it grows larger.

Project: Investigating Fish Scales

Can you estimate the total number of scales covering the skin of your fish? Count the scales in a square inch (an area 1 inch by 1 inch) (or use metric units). Then measure the side of the fish and figure out how many square inches or square centimeters of scales there are. Does your fish have more than a thousand scales?

Remove fish scales from several different parts of your fish. Look at one with a microscope under low power and count the number of lines on the scale. Are there more than ten? Examine other scales to see if they all have the same number of lines.

Try to get scales from different kinds of fish. Go to the fish market and ask the workers to save scales for you when they clean fish. Do big fish always have larger scales than smaller fish? Make drawings to show the size and shape of different fish scales.

INTERNAL ORGANS

The main internal organs are contained in an abdominal cavity on the bottom of the fish, behind the head. **Working under adult supervision**, open the abdominal cavity as follows:

1. Insert the scissors into the *anus* (Figure 14A).

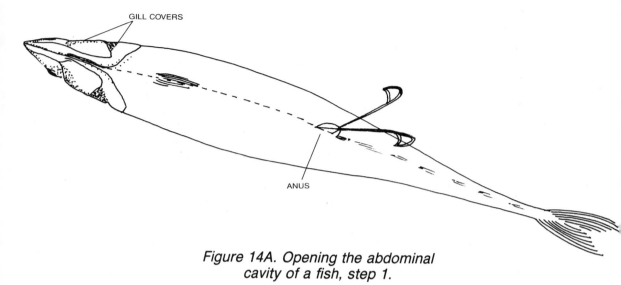

GILL COVERS

ANUS

*Figure 14A. Opening the abdominal
cavity of a fish, step 1.*

2. Cut a slit through the skin and flesh to the *gill covers.* (Figure 14A). Do not damage the organs inside.
3. Cut off flesh on one side from the anus to the gills (Figure 14B).

To remove the *organs* from the cavity (Figure 14C):

1. Snip off the end of the intestine where it leaves the belly at the anus.
2. Cut away the thin tissues holding the organs to the inside walls of the cavity.
3. Lift out the organs from the back end with your fingers. Cut more tissues as necessary.
4. Cut the tube leading from the stomach to the mouth.
5. Remove the organs.

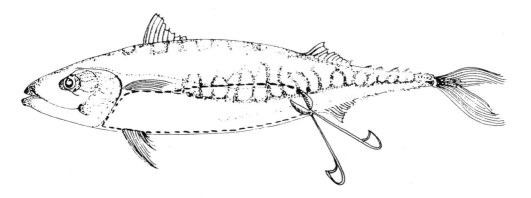

*Figure 14B. Opening the abdominal
cavity of a fish, step 2.*

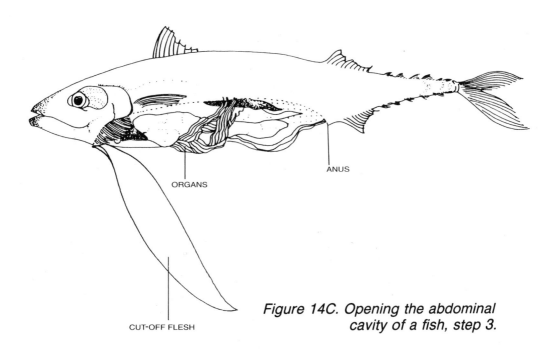

ORGANS

ANUS

CUT-OFF FLESH

*Figure 14C. Opening the abdominal
cavity of a fish, step 3.*

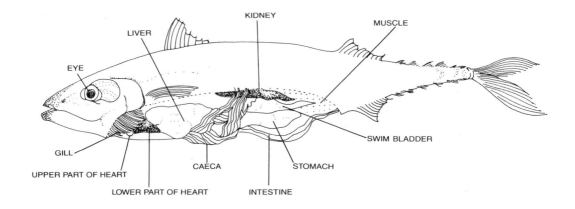

Figure 15. Internal organs of a mackerel

Refer to Figure 15 for the following.

You will find the *heart* toward the front near the jaw. The bottom part of the heart is dark red and has a triangular shape. Attached on top is a smaller white part. Try to remove the heart in one piece. Then hold it between the forceps and cut the lower part in half. Can you find a space inside where the blood is pumped? If you see any blood, pick up a few drops in the pipette. You might be able to see the blood cells with a microscope.

The size of the stomach depends on how much food it contains. Cut open the *stomach* so that you can find out what the mackerel might have eaten. Is the *intestine* long and winding like yours? The *liver* is light tan and has three parts. The spaghetti-like strings are *caeca*, which are thought to aid in digestion.

The fish uses its *swim bladder* for floating. When air is taken into the bladder, the fish becomes lighter and floats. Sharks have no swim bladders and must continue swimming or sink.

The two *kidneys* are attached to the top of the cavity. They are dark red. The kidneys remove liquid wastes from the body.

Still working under supervision, use your scalpel to remove an *eye* from the skull. Push the point of the scalpel between the outside of the eye and the skull socket. Then cut around in a circle until the eyeball comes out. Cut open the eye with the scissors. The round shape of the eye is maintained by the watery liquid that oozes out. Look for a small, round "bead"—the lens. The lens focuses the light to create an image on the inside of the eyeball. Place the lens on some small words cut from a newspaper or magazine. The lens will act like a magnifying glass to enlarge the letters. How do the fish's eyes compare with the squid's or lobster's?

The part of the fish that you eat is *muscle*. Most of the body and tail of a fish are composed of muscles, since rapid swimming requires great effort. Use the scalpel or scissors to remove the skin from part of the body. Cut off some of the muscles so you can see how they are arranged in layers.

Expose the fish's skeleton by cutting off all of the muscles from the body and tail. This is how a fillet is made: the muscles are removed from the bones for cooking. If you have trouble getting the small bones clean, cut off the head and boil the backbone and tail for a few minutes. An old toothbrush can be used to remove the last pieces of cooked muscle from the bones.

The backbone is made up of a series of small bones called vertebrae. You an see the hole for the nerve cord in the end of the vertebra that was attached to the head. Notice the movable joints between the vertebrae. Bend the backbone from side to side the way a fish does when it swims. Can the backbone bend more from side to side or up and down? Why?

A fish skeleton has many ribs attached to its backbone, but these usually come off when the muscle is removed. The spines attached to the vertebrae are not ribs.

The skull is made of many different bones. To clean them, cook the skull in boiling water for 5 to 10 minutes. Most of the bones will fall apart. After the bones have dried, you can glue them to a board for display.

Meat markets sell bones,
all kinds of organs, and sometimes
unusual parts of animals.

DISSECTING A BEEF
ANKLE JOINT

You should be able to get an ankle joint from the butcher in a supermarket. Since bones are sold for dogs, it may cost a few dollars.

Are your surprised by the weight of the joint? Bones must be strong and heavy to provide the strength needed to support an animal's body. The lower leg bone, or *tibia*, has been cut off a short way above the ankle. Use the paring knife to try to cut into the outside of the bone. **Be sure you are working under adult supervision when using the knife or scalpel.** It is probably so hard that you will be unable to remove even a small slice.

Refer to Figures 16 and 17 for the ankle anatomy.

FROM MARROW TO TENDONS

Have an adult cut the bone in half with a hacksaw. What do you notice about the inside of the bone? Instead of being hard, it is filled with mushy, red tissue. Hollow bones are lighter than solid bones and provide almost as much strength. The bones of birds are even thinner; light weight is important for flight.

The red tissue inside the bone is *marrow*, which fills the insides of most of the long bones in your arms and legs. Push the

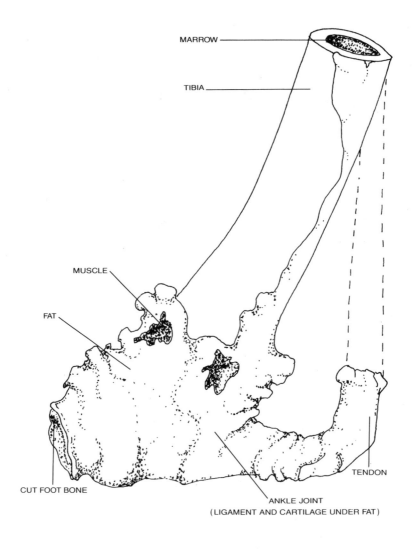

MARROW

TIBIA

MUSCLE

FAT

CUT FOOT BONE

ANKLE JOINT
(LIGAMENT AND CARTILAGE UNDER FAT)

TENDON

*Figure 16. Beef ankle joint
from a butcher shop*

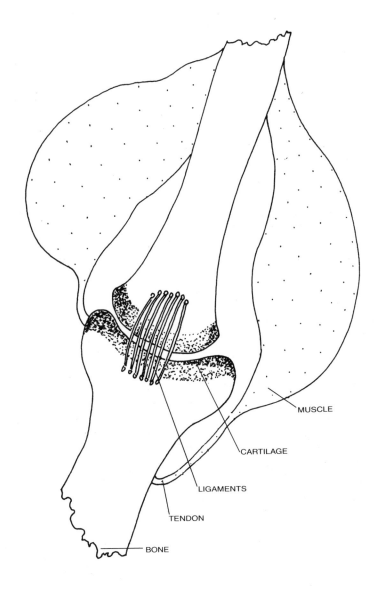

MUSCLE

CARTILAGE

LIGAMENTS

TENDON

BONE

Figure 17. Diagram of a joint

teasing needle down into the marrow. Your needle will sink in easily until it strikes the lower end of the bone. Spoon out some of the marrow with the paring knife blade. Smell it. It is this tangy flavor that causes dogs to chew on bare bones in an attempt to get at the marrow inside. Soup is sometimes flavored by boiling bones to extract the flavorful marrow.

If you have a microscope, you can look at some marrow with it. Spread a tiny bit out on a microscope slide in a thin layer. What do you see?

Blood cells are produced by the bone marrow, which is why it is red. Human red blood cells wear out in several months and must be replaced with new ones. Right now your bone marrow is producing more than 70 million red cells—every second!

There is a lot of white tissue covering the outside of the ankle joint. You probably know that much of this is *fat*. Find a soft spot and cut away a blob of fat. Notice how greasy it is. Some supermarkets sell fat as suet to feed birds in the wintertime. Your body uses fat to store food and as insulation against the cold. The fat around the ankle joint protects it from being injured when bumped.

You will also see a few small spots of red tissue. These are parts of *muscles*. Why has the butcher cut off most of the muscles that were once above the ankle joint? Muscles, of course, are the meat you eat in steak, roast beef, and hamburgers.

Attached to the back of the heel you will find another kind of white tissue. Try to cut it with the scalpel. Is it soft like fat? This tough cord is called a *tendon*. Tendons connect muscles to bones. Pull on the tendon to make the foot bones move down. If muscles attached directly to bones, the joints would be too fat and stiff.

The heel tendon is named the Achilles tendon. Feel it in your own foot above the heel. Achilles was a famous warrior in Greek mythology. When Achilles was a baby, his mother dipped him into a magic potion that would protect him from being injured with spears and swords. As a soldier, Achilles won many battles because no weapons could harm him. But one day he was killed by an arrow that struck him in the tendon on the back of his heel.

This spot was unprotected because it was there that his mother's fingers had held him in the potion.

EXAMINING THE JOINT

Now you are ready to expose the inside of the ankle joint. You will find it hard to remove all of the tissues. The paring knife and dissecting scissors will be most useful. It is helpful to have someone else hold the joint in different positions while you continue to cut away.

As you get down to the joint, you will find tough bands connecting the tibia to the *foot bones*. These are *ligaments*. They hold the bones together at a joint but still allow movement in the proper direction. The *ankle joint* moves mostly up and down but very little from side to side. Move the joint as much as you can in various directions.

If one of *your* joints moves too far to the side, you get a sprain. When you sprain your ankle, the tendons get stretched too much. A bad sprain can cause torn ligaments. This occurs when some of the ligament pulls away from the bone. Football players often suffer torn ligaments in knee joints.

Now cut away all of the ligaments and separate the tibia from the foot bones of the beef ankle joint. It will not be easy; the ankle joints must be strong enough to hold up the 1,500-pound (680-kg) steer when it runs. **Be careful.** The beauty of the exposed joint will be your reward for the hard dissection work. The smooth coating on the ends of the bones is *cartilage*. There must be very little friction where the bones of a joint rub together. The cartilage is "greased" by oils from the surrounding fat.

Cartilage is the same flexible tissue that shapes your nose and ears. Feel it. Cartilage bands also connect the front of your ribs to the breastbone. The rib cage must stretch and expand everytime you take a breath.

If you have a dog, he will certainly enjoy finishing the dissection of the ankle bones. Unlike you, he is just interested in eating rather than in studying the different tissues that permit the joint to function properly.

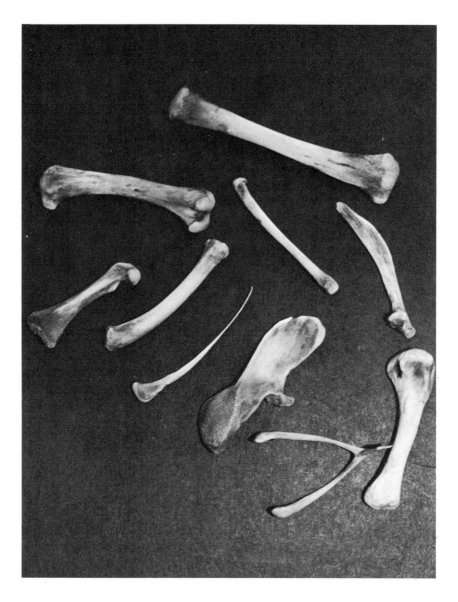

An assortment of bones from
a store-bought chicken

DISSECTING A CHICKEN
LEG AND WING

Probably you already have dissected chicken wings and legs many times with your teeth. When eating fried chicken, you bite off the skin and meat from the bones.

It is easy to get chicken legs and wings; all supermarkets sell fresh chicken parts. If you buy a "quartered" chicken, you will have two wings and two legs for dissection.

Work under adult supervision when using the scalpel or knife and be sure to wash your hands when you finish the activities.

EXTERNAL EXAMINATION

If the leg is still attached to the body, use the scalpel or paring knife to cut it off at the hip. Move the leg around and slice into the *hip joint* until the leg can be separated. See Figure 18.

Pick up the leg and bend it at the *knee joint*. The section above the knee is the *thigh*. The longer leg section below the knee is known as the *drumstick*. Why? Unlike your leg, a chicken leg does not end at its ankle, but the end of the leg you are examining is the *ankle joint*. Since one of the *foot bones* has become lengthened, the bird always walks on tiptoes. A long leg enables the chicken to run faster, which is important since it cannot escape danger by flying.

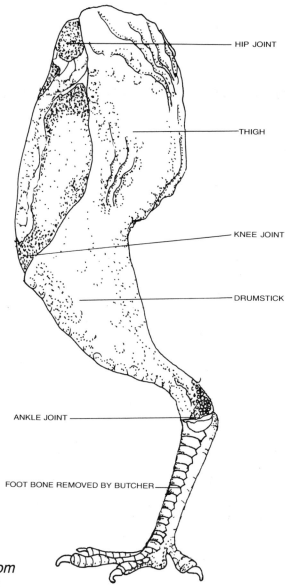

HIP JOINT

THIGH

KNEE JOINT

DRUMSTICK

ANKLE JOINT

FOOT BONE REMOVED BY BUTCHER

Figure 18.
Chicken leg from
a supermarket

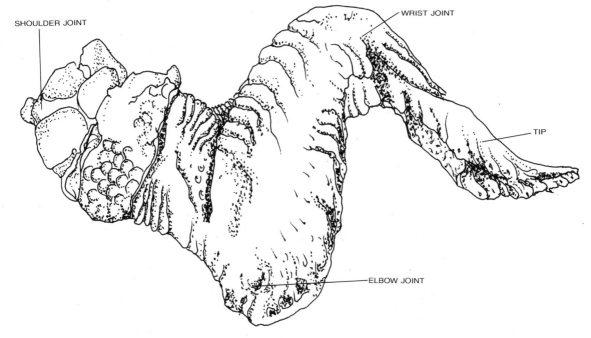

SHOULDER JOINT

WRIST JOINT

TIP

ELBOW JOINT

*Figure 19. Chicken wing
from a supermarket*

There is no skin on the inside of the thigh where it attached to the bird's body. Is your chicken leg a left one or a right one? Why is the bottom leg section cut off before the chicken is sold for food?

Refer to Figure 19 for the following.

Cut the wing from the body at the *shoulder joint*. How many sections does the wing have? There are three main parts, but if you wiggle the thin *tip*, you can feel two more small bones at the end. The wing has an *elbow joint* and *wrist joint* as you have. Many of the bird's "hand" bones are missing, however.

Bend your own arms into the shape of folded wings. Is your chicken wing from the right side or the left side? Why is there a flap of skin stretched across the wing at the elbow?

How is chicken skin different from your own skin? Pinch the skin on your arm so you can feel its thickness. Is chicken skin thinner than yours? Look at all the bumps covering the chicken skin. The bumps are arranged in a pattern. When the chicken was alive, there was a feather on each bump. Much of *your* skin is covered with hair growing from tiny holes. Examine the chicken skin under a microscope.

INTERNAL STRUCTURES

Now remove the skin from the leg (see Figure 20). To do this,

1. Use scissors to cut a slit in the skin from the knee to the ankle.
2. Pull the skin from the meat with your hands.
3. Cut the skin away in places where it does not separate.

The "meat" on the leg is muscle. There are two main muscles on each part of the leg. Muscles are almost always found in pairs, since a muscle can only pull in one direction. When your lower leg moves, one muscle pulls it backward, and another muscle pulls it forward. The large muscles on the drumstick move the lowest leg section. The muscle is attached to the bone it moves with a *tendon*, a tough white cord. You should be able to find the tendons just above the *ankle joint*, one on either side for each muscle. The rounded end on the upper leg bone fits into a socket in the hip bones. This is the *hip joint.*

  See if you can cut the skin from the wing. Look for pairs of muscles and the tendons attached to them. (If you like to draw, you might try to redraw Figure 19 to show the muscles.) Notice how the muscles are covered by a clear, thin tissue. The wing muscles are not as dark as the leg muscles. When eating a chicken, you know that the cooked wing has "light meat" and the leg is "dark meat." Dark meat is colored by the greater supply of blood necessary for the muscle to function actively; the muscles in the weak chicken wings have less blood. Why do chicken wings need less blood than the legs?

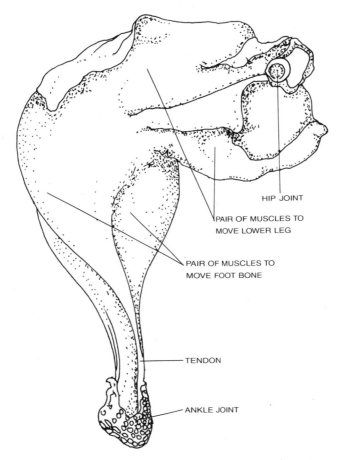

HIP JOINT

PAIR OF MUSCLES TO
MOVE LOWER LEG

PAIR OF MUSCLES TO
MOVE FOOT BONE

TENDON

ANKLE JOINT

*Figure 20. Chicken leg
with the skin removed*

Can you find yellow or white tissue attached to the inside of
the skin and in cracks between some muscles? This is fat.

The bones of the wings and legs are under the muscles. In
order to investigate the bones you will need to clean them as
follows:

should be soft enough to bend into a knot. In Chapter 4 there is a similar project with a seashell. Do you think the chicken bone contains calcium? Test other substances in vinegar to see if fizzing occurs. Try such things as small rocks, an eggshell, a piece of your fingernail, a tooth, a piece of chalk, and fish bones. Do all of these contain calcium too?

10

DISSECTING A
WHOLE CHICKEN

When your grandmother was young, grocery stores sold only whole chickens: birds complete with head, feet, and guts. The only things removed were the feathers. There were no fresh chicken parts or frozen chickens.

Today it is still possible to buy whole chickens at some poultry markets. The customer is shown a live chicken for approval. Then, after the bird is killed, its feathers are removed with a special machine. The body of the chicken is still warm when it is carried out of the market.

The butcher at a poultry market might give you the internal organs from a chicken. Usually chickens are cleaned after being killed, and the insides are discarded in a "slop barrel."

It might be possible to obtain a whole chicken from a poultry farm. Then you would have to remove the feathers yourself. If you can't get such a chicken, you can still do many of the activities in the chapter using a "whole" chicken from the supermarket. Just make sure it comes with the little paper sack full of internal organs.

EXTERNAL EXAMINATION

If your chicken came from a poultry market, you will see that the bird's neck has been cut. This is how the chicken was killed.

This store sells live chickens and ducks. Most people pick out a live bird, which the butcher then dresses. The store is located near Chinatown and Little Italy, in New York City.

Chickens have evolved from birds that were good fliers. For this reason, a chicken's body is designed for flying even though it can fly for only short distances. Its wings are too small to support such a heavy body in the air. A light head is necessary for efficient flight. Look at the size of your chicken's head. One thing that makes birds' heads light is their small size.

Open the mouth and look for teeth. You won't find any. Instead of heavy teeth, birds have strong but light beaks. Does the chicken have a tongue? Can you find any ears? A chicken does have inside ear parts much like yours, but it has no external ears. Would outside ears hinder a flying bird? Does a chicken have nostrils? Are there any eyelids? Where is the comb?

Stretch out a wing (Figure 23) and move it as if the bird were flying. Notice how the flap of *skin* in front of the elbow increases the size of the wing's surface. Of course, when covered with long feathers, the wing was much larger. Are any small feathers still left on the wing tip?

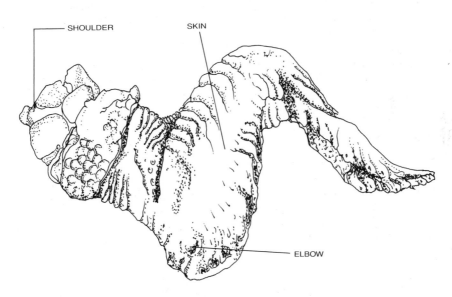

Figure 23. External parts of a chicken wing with feathers removed

–73

The three main sections of the wing move at the joints where the bones connect. The *shoulder* has a ball-and-socket joint, which allows free movement in any direction. At the *elbow* is a hinge joint, which moves most easiy in just one way, back and forth. Where does your body have hinge joints?

A chicken leg also has three main parts. The upper section is embedded under the skin on the body and is less noticeable than the lower two sections. The bottom leg section with the toes is removed from chickens sold in supermarkets.

Look at the chicken's feet. How are they designed for scratching soil in search of food?

Project: Making Chicken Tracks

You can make chicken tracks with a chicken leg and a rubber stamp ink pad. First, cut off the lower leg section at the joint with your scissors or paring knife. Spread out the toes and press the foot on the ink pad. Then stamp the foot on a piece of paper. You could use both feet to stamp tracks that look like a chicken had walked across the paper. Does the right foot track differ from the track made by the left foot?

Use thick poster paint to make a track with *your* foot. Other than size, how is your foot different from a chicken's? How do the chicken's skinny toes and long toenails help it to find food? Why can a chicken hold onto a perch better than you can?

INTERNAL STRUCTURE

The internal organs of a chicken are inside a large abdominal cavity (see Figure 24). When chickens and turkeys are roasted, the cavity is usually filled with bread crumbs for stuffing.

Working under adult supervision, remove the organs from the chicken (you won't need to do this if your chicken comes from the supermarket):

1. Place the chicken on its back.
2. Start at the anus and cut open the skin across the breast to the base of the neck.

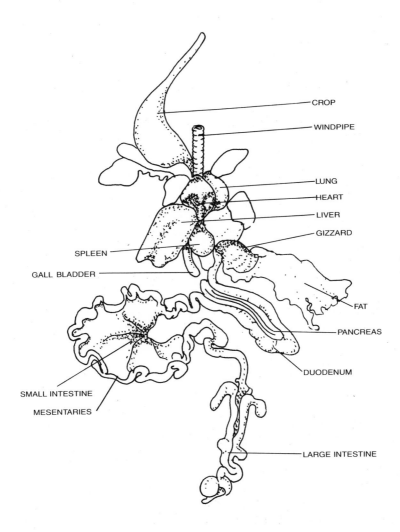

CROP

WINDPIPE

LUNG

HEART

LIVER

GIZZARD

SPLEEN

GALL BLADDER

FAT

PANCREAS

DUODENUM

SMALL INTESTINE

MESENTARIES

LARGE INTESTINE

Figure 24. Internal organs of a chicken

3. Remove the skin from the breast area.
4. Cut off the meat from the top of the breastbone.
5. Cut the breastbone away from the ribs.
6. Snip off the end of the intestine at the anus.
7. Cut away the thin tissues holding the organs to the inside walls of the cavity.
8. Lift out the organs from the hind end while cutting more tissues as necessary.
9. Cut the tubes leading up the neck.
10. Remove the organs.

Most of the organs are part of the chicken's digestive system. The chicken has the same organs as you have plus a few extra ones. Corn and other grains are harder to digest than the hot-dogs and spaghetti you eat. (Store-bought chickens usually have only the heart, liver, and gizzard in a separate package).

At the base of the neck is a large, thin bag called the *crop*. After food is swallowed, it is stored here until it can enter the digestive system. The crop may have been cut open when the organs were removed. Inside the crop you will find a yellow-brown mush: the last grain eaten by the chicken. You do not have a crop; the food you swallow goes right into your stomach.

The *gizzard* is another organ that you do not have. The strong muscular walls of the gizzard grind up the grain into smaller par-ticles that can be digested more easily. A chicken cannot chew its food as you do because it has no teeth.

Find the gizzard. It is a hard, red lump covered by thick layers of yellowish *fat*. To dissect the gizzard, (1) strip the fat from the gizzard, (2) slice the gizzard open and turn it inside out; (3) rinse off the crushed grain in running water, and (4) peel away the rough inner lining with your fingers. Rub the lining on your hand. Try to tear it. Gizzards, with the inner lining removed, can be cooked and added to giblet gravy.

Food from the gizzard goes into the *duodenum*. From there it passes into the *small intestine* and then to the *large intestine*. The intestines are held in place by thin tissues called *mesenteries*. Carefully cut the mensenteries so that you can stretch out the intestines in a straight line. Measure the total length of the chicken's intestines. *Your* intestines may be more than 7 feet (2 m) long!

The chemicals needed to digest food are secreted by glands. The largest gland is the *liver*. It is dark red and has three main parts, or lobes. Cut the liver open. Is it hollow inside? Some people eat chicken liver after it is fried in butter and mushrooms. Do you think you would like chicken liver?

Another gland, the *pancreas*, is located between the two loops of the duodenum. The *gall bladder* is easy to find because of its green color. The rounded *spleen* is dark red.

Remove the *heart* by cutting off the blood vessels on top. Notice that the heart is enclosed in a transparent sac. Is a chicken heart shaped like a Valentine heart? Cut the heart open and try to find the two main chambers inside.

The dissection of a beef heart is described in Chapter 11. Even though a chicken heart is small, you should be able to find some of the same parts shown in the beef heart.

The heart, too, is often used to flavor giblet gravy.

The *wind pipe*, or air tube leads from the mouth to the *lungs*. Try to pinch shut the air tube with your finger or the forceps. Stiff rings of cartilage prevent the tube from collapsing.

Project: Making a Chicken Skeleton
Cook all the parts of the chicken (except the insides) in slowly boiling water for 1½ to 2 hours. After cooling, scrape the meat from the bones with your fingernails or a scalpel. Wash off the bones and allow them to dry overnight on a newspaper.

How many bones do you count? If it's near 120, you're about right. Glue the bones together on a piece of cardboard in the right order. Label any that you can identify.

Wash your hands and the surface you worked on when you are through working with the chicken, as chickens sometimes carry diseases that can make you sick.

11

DISSECTING A BEEF HEART

Beef hearts can be obtained from slaughterhouses or scientific supply companies. To find a slaughterhouse, look in the Yellow Pages of your telephone book under "Abattoir." Large slaughter-houses kill beef cattle for steaks and roasts. Calves, sheep, and pigs might be slaughtered at smaller abattoirs for veal, lamb, and pork, respectively.

The heart from any animal slaughtered for meat makes a good specimen for dissecting. When ordering a heart, request that it be removed with the blood vessels intact. It should cost you only a few dollars.

If you can't get one of these hearts, you can substitute a chicken heart. However, a chicken heart is much smaller and harder to work with.

You can keep the heart fresh for a few days in your refrigerator. Or store it in the freezer until you are ready to dissect.

EXTERNAL EXAMINATION

Refer to Figure 25 during the external examination of the beef heart.

Your own heart is about the same size as your fist. Compare the size of the heart you will dissect with that of your fist. Is the animal heart bigger or smaller than your heart? Would you say a heart is square, round, or triangular?

-79

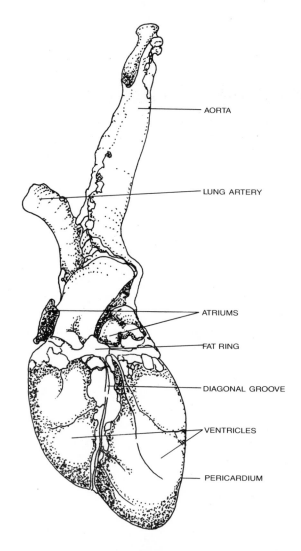

AORTA

LUNG ARTERY

ATRIUMS

FAT RING

DIAGONAL GROOVE

VENTRICLES

PERICARDIUM

*Figure 25. External parts
of a beef heart*

The heart is enclosed in a thin sac, the *pericardium*. When you rub the heart with a finger, you should see the clear membrane moving around.

Squeeze the heart. It is so solid that you would not expect to find any spaces inside. But there are two large chambers (the *ventricles*) inside the heart, as you will see later when you dissect it. The diagonal groove shows the division between the two lower chambers.

The heart is a pump, actually two pumps. Thick, muscular walls are required to pump blood continuously. All of the red tissue you see beneath the pericardium is muscle. A *fat ring* circles the heart at its widest part near the top.

Two small chambers are on top of the heart. These are the atriums. They are the dark-red structures just above the fat ring.

The largest tube leaving the top of the heart is the *aorta*. It carries blood from the heart to all parts of the body except the lungs. The smaller artery is the *lung artery*. Veins are blood vessels that return blood to the heart. The walls of veins are not as thick as artery walls. The smaller blood vessels on the surface of the heart supply the heart muscle with blood.

Sometimes part of a lung is cut off with the heart. Lung tissue is spongy and almost white. The lungs are located close to the heart since the blood must transport oxygen throughout the body.

INTERNAL STRUCTURE

Working under supervision dissect the heart:

1. Cut off the pericardium from outside the heart.
2. Position the heart so that the diagonal groove is facing upward.
3. Cut through the heart from the tip toward the top, as shown in Figure 26A.
4. Remove a section of the muscle wall as shown in Figure 26B.

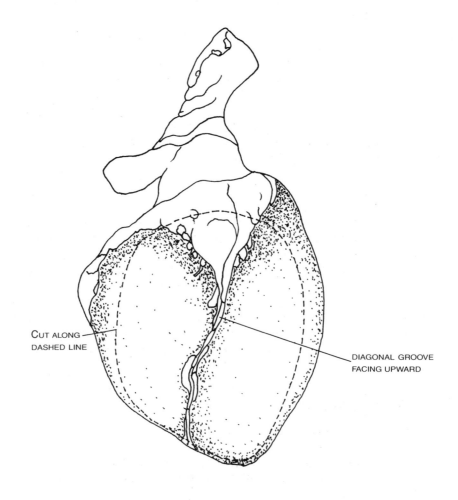

CUT ALONG
DASHED LINE

DIAGONAL GROOVE
FACING UPWARD

Figure 26A.
Opening a beef heart, step 1.

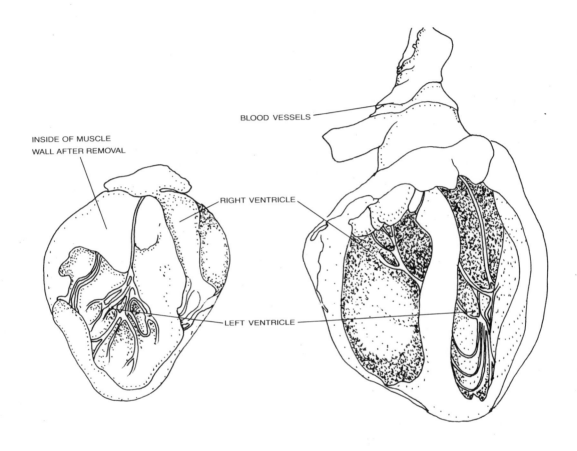

INSIDE OF MUSCLE
WALL AFTER REMOVAL

BLOOD VESSELS

RIGHT VENTRICLE

LEFT VENTRICLE

*Figure 26B.
Opening a beef heart, step 2.*

Refer to Figure 27 for the following activites and observations.

You have exposed the two main chambers of the heart, the ventricles. The ventricles are separated by the *septum* wall between them. Are both ventricles the same size? The larger one is the *left ventricle*, which pumps blood through the body; the smaller *right ventricle* pumps blood only to the lungs. Why is the right ventricle smaller? When the heart pumps, the ventricle walls contract, forcing blood out through the arteries.

Toward the top of the ventricles you will see a number of thin white strings. These are parts of the *cuspid valves*, which keep blood flowing in the proper direction. The strings are attached to thin flaps that snap shut every time the heart beats. Poke around under the strings with the handle of the teasing needle until you discover one of the little flaps. Cuspid valves like these in your own heart make the heart noises you can hear after exercising.

Sometimes blood clots are found lodged in the valves. If the animal heart has any, pick them out with the forceps.

Blood enters the ventricles from the small upper chambers, the *left atrium* and the *right atrium.* Take the handle of the teasing needle and try to push it from the top of a ventricle into the atrium. Why do you think the walls of the atria are so much thinner than the ventricle walls?

While probing for an atrium, you may have pushed your teasing needle up into the *aorta* or *lung artery*, through which blood leaves the ventricles. Push the handle of the needle all the way through the artery until it comes out of the top. Another way to find where an artery or vein comes from is to push the handle into the top of it and watch where it comes out in the heart. Veins lead into the atria, but these may have been cut off before you got the heart.

Another kind of valve in the arteries prevents blood from running back into the heart between beats. Again using the handle of the teasing needle as a probe, look for thin pockets at the base of an artery where it joins to the top of the ventricle. The pocketlike valves are shaped like half-moons, hence the name *semilunar* valves.

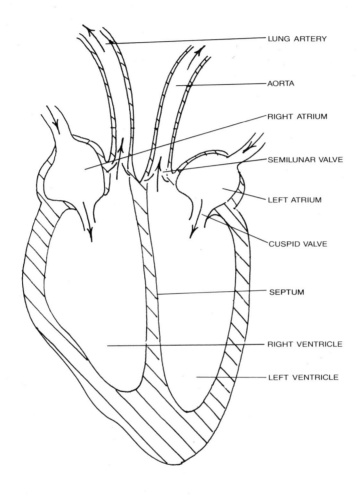

LUNG ARTERY

AORTA

RIGHT ATRIUM

SEMILUNAR VALVE

LEFT ATRIUM

CUSPID VALVE

SEPTUM

RIGHT VENTRICLE

LEFT VENTRICLE

*Figure 27. Diagram of
a beef heart*

Project: Making a Model Valve

Figure 28 shows how to make a model to show the action of the heart valves. You will need a paper tube from inside a paper towel roll. After the valve is made, point the tube upward and blow out. Then try to breathe through the tube. The flap on top should close up and stop the air.

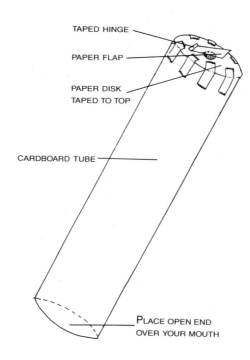

TAPED HINGE

PAPER FLAP

PAPER DISK
TAPED TO TOP

CARDBOARD TUBE

PLACE OPEN END
OVER YOUR MOUTH

*Figure 28. How to make
a model valve*

12

DISSECTING A CALF EYE

Animal eyes can be obtained from slaughterhouses where calves and steers are butchered.

EXTERNAL EXAMINATION

Refer to Figure 29 for the following observations and procedures.

When you look at an eye removed from an animal's head, you can understand why the term "eyeball" is used. The eye is almost perfectly round, like a ball.

The clear part of the *eyeball* in front is the *cornea.* Try to scratch the cornea with your fingernail. The cornea is so tough that you will find it difficult to cut with your scissors when you perform the dissection.

Look at the cornea from the side. Do you notice how it has a slight bulge? Here is how you can feel the bulge in the cornea on *your* eye: Shut one eye and touch your closed eyelid. Then move the open eye up and down. You should be able to feel the cornea moving under the eyelid.

If you want a surprise, look at the calf eye cornea and squeeze the eyeball hard. What happens to the color inside?

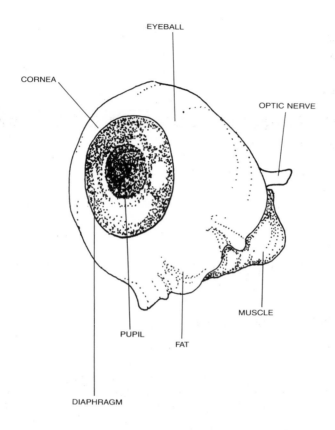

EYEBALL

CORNEA

OPTIC NERVE

MUSCLE

PUPIL

FAT

DIAPHRAGM

Figure 29. External parts of a calf eye

Look into the cornea. What can you see inside the eye? The dark ring is the iris, or *diaphragm*, which controls the amount of light that enters the eye. The diaphragms of your eyes are what give them their color: blue, brown, or hazel.

The dark circle in the center of the diaphragm is the *pupil*. The pupil is merely a space in the diaphragm. The hole looks black because the inside of the eye is dark. The size of the pupil changes automatically; it is large in dim light and smaller in bright sunlight.

Clinging to the outside of the eyeball in back are red and white tissues. The soft white material is *fat*, which cushions the eyeball from its bony socket in the skull. The red tissue is the remains of *muscles* that moved the eyeball around.

Also in back you should find a white cord that looks like a short piece of spaghetti. This is the *optic nerve*, which carries messages from the eye to the brain.

INTERNAL STRUCTURES

You will need a saucer in the procedure to open the eye:

1. Hold the eye firmly with one hand and cut a hole with the point of the scalpel in the eyeball at the outer edge of the cornea (Figure 30A).

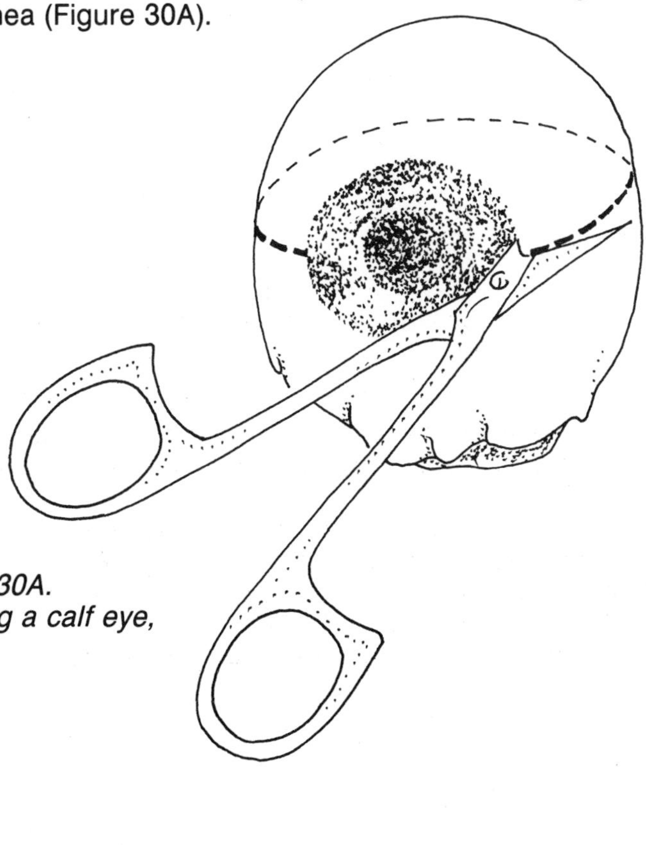

Figure 30A.
Opening a calf eye,
step 1.

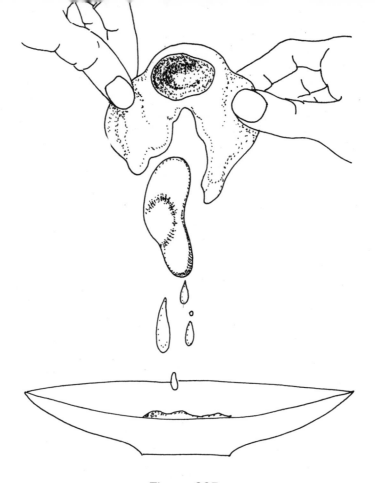

Figure 30B.
Opening a calf eye, step 2.

2. Insert one point of the scissors into the hole and cut all the way around the back of the eyeball to the other side of the cornea (Figure 30A).
3. Hold the eye with both hands over the saucer with the cut side of the eyeball on the bottom (Figure 30B).
4. Lift up on each side of the eyeball until the inside drops out into the saucer (Figure 30B).
   Be sure you are working under adult supervision.

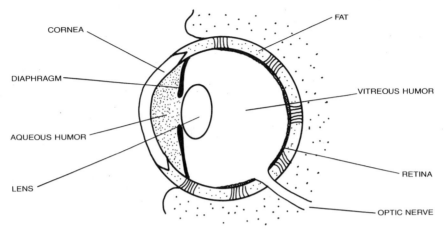

Figure 31.
Diagram of a calf eye

Refer to Figure 31 for the following activities and observations.

You might first notice a blob of clear jellylike substance. This is the *vitreous humor* that fills most of the eyeball. What has happened to the eyeball after the vitreous humor came out?

On top of the vitreous humor is the *lens.* It might remind you of a flattened plastic bead. The lens acts as a magnifying glass to focus light from things you see. If the eye is not fresh, the lens will have become cloudy.

Use the teasing needle to tease the lens from the vitreous humor onto a piece of newspaper with small words. Look at the print through the lens. What do you see?

The lens can be preserved in a bottle of alcohol.

The black veil that you might find around the lens is the *diaphragm.*

There is a small amount of watery liquid inside the eye just behind the *cornea.* This is the *aqueous humor.* If you can see any, pick up a few drops in the pipette. Is the aqueous humor perfectly clear?

Now turn the eyeball inside out. The back of the eye is covered with a beautiful bluish film, the *retina*. The image of what you see forms on the retina, which can be compared to the film in a camera. Look on the inside of the eyeball opposite the *optic nerve*. What do you notice about the retina at that spot?

With your scissors, cut around the outer edge of the *cornea* until it can be separated from the rest of the eyeball. Look through the cornea. In live animals, the cornea "window" is perfectly clear.

FOR FURTHER
READING

Berman, William. *How to Dissect.* 4th ed. New York: Arco, 1985.

Bliss, Dorothy. *Shrimp, Lobster, and Crabs: Their Fascinating Life Story.* Piscataway, N.J.: New Century, 1982.

Buchsbaum, Ralph et al. *Animals without Backbones.* 3rd ed. Chicago: University of Chicago Press, 1987.

Bunting, Eve. *The Giant Squid.* New York: Messner, 1981.

Goldsmith, Ilse. *Human Anatomy for Children.* New York: Dover, 1969.

Headstrom, Richard. *All about Lobsters, Crabs, Shrimps, and Their Relatives.* New York: Dover, 1985.

Zim, Herbert. *What's Inside of Animals.* New York: Morrow, 1953.

INDEX